Grades 5-8

Focus On

MIDDLE SCHOOL

Laboratory Notebook

3rd Edition

Rebecca W. Keller, PhD

Real Science-4-Kids

Illustrations: Janet Moneymake

Copyright © 2019 Gravitas Publications Inc.

All rights reserved. No part of this publication may be reproduced, stored in a retrieval system, or transmitted, in any form or by any means, electronic, mechanical, photocopying, recording, or otherwise, without prior written permission from the publisher. No part of this book may be reproduced in any manner whatsoever without written permission.

Focus On Middle School Geology Laboratory Notebook—3rd Edition
ISBN 978-1-941181-55-3

Published by Gravitas Publications Inc.
www.gravitaspublications.com
www.realscience4kids.com

Keeping a Laboratory Notebook

A laboratory notebook is essential for the experimental scientist. In this type of notebook, the results of all the experiments are kept together along with comments and any additional information that is gathered. For this curriculum, you should use this workbook as your laboratory notebook and record your experimental observations and conclusions directly on its pages, just as a real scientist would.

The experimental section for each chapter is pre-written. The exact format of a notebook may vary among scientists, but all experiments written in a laboratory notebook have certain essential parts. For each experiment, a descriptive but short *Title* is written at the top of the page along with the *Date* the experiment is performed. Below the title, an *Objective* and a *Hypothesis* are written. The objective is a short statement that tells something about why you are doing the experiment, and the hypothesis is the predicted outcome. Next, a *Materials List* is written. The materials should be gathered before the experiment is started.

Following the *Materials List,* the *Experiment* is written. The sequence of steps for the experiment is written beforehand, and any changes should be noted during the experiment. All of the details of the experiment are written in this section. All information that might be of some importance is included. For example, if you are to measure 236 ml (1 cup) of water for an experiment but you actually measured 300 ml (1 1/4 cup), this should be recorded. It is hard sometimes to predict the way in which even small variations in an experiment will affect the outcome, and it is easier to track down a problem if all of the information is recorded.

The next section is the *Results* section. Here you will record your experimental observations. It is extremely important that you be honest about what is observed. For example, if the experimental instructions say that a solution will turn yellow, but your solution turned blue, you must record blue. You may have done the experiment incorrectly, or you might have discovered a new and interesting result, but either way, it is very important that your observations be honestly recorded.

Finally, the *Conclusions* should be written. Here you will explain what the observations may mean. You should try to write only valid conclusions. It is important to learn to think about what the data actually show and what cannot be concluded from the experiment.

Contents

Contents

Experiment 1

Observing Your World

Photo Credit: US Geological Survey/by Lyn Topinka

Introduction

Do you think if you go outside and look carefully, you will see things you haven't noticed before? Try it!

I. Think About It

❶ What features do you think you will see when you go outside?

❷ What do you think the dirt in your yard or in the park looks like?

❸ What geological features do you see when you go to the grocery store?

❹ What man-made features do you see when you go to the grocery store?

❺ In what ways does the weather change where you live? Do you think any changes in the weather affect the landscape?

❻ Do you think places such as a city or a rural area have a history? Why or why not?

II. Experiment 1: Observing Your World

Date ____________

Objective

____

____

Hypothesis

____

____

____

Materials

pencil, pen, colored pencils
small jar
trowel or spoon
binoculars (optional)

EXPERIMENT

1. Step outside your front or back door and walk until your feet are on some type of ground (dirt, grass, or concrete).

2. Observe where you are. Are you in a city? Are you in the country? Use the space below to draw or write what you see.

❸ Observe any geological features near you. Do you see mountains? Do you see lakes or rivers? Do you see the ocean? Do you see other geological features? Record what you are observing.

❹ Use the trowel to collect a small sample of dirt. (If you live in the city, walk to a park or some other place where you can collect a dirt sample.)

❺ Observe the dirt sample. Is it light in color? Dark? Does it contain small rocks? Large rocks? Does it have any organic matter (living things, such as grass or bugs)? Record what you observe.

❻ Observe any man-made structures. Do you see buildings? Roads? Bridges? Other man-made structures? Record what you see.

❼ Observe any dynamic features in your area including the weather. Do you get earthquakes? Do you live near a volcano? Does it rain frequently, or do you get very little rain where you live? Do you have tornadoes, hurricanes, or severe snow storms?

8. Think about the area in which you live. What is its history? How long has it looked the way it looks today? If you are in a city, how long has the city been there? What do you think it looked like before there were buildings, roads, or other structures? Write your observations below.

Results

Assemble your data in the chart below.

Data Describing Where I Live	
Geological Features	
Soil Type	
Man-made Structures	
Dynamic Processes	
Weather	
History	

III. Conclusions

Review the observations you made during this experiment. What did you observe that you had not noticed before? Do you think that taking the time to look at things carefully makes a difference in what you observe? Why or why not?

IV. Why?

The first step towards understanding the world around you is to observe it. Before you can know what the world looks like, how it changes, and how the different parts work together, you have to go outside and observe buildings, rocks, the soil, clouds, sunlight, plants, and animals. You need to observe the weather, how it changes, and how it affects the landscape and living things. Observations made over time are important for noticing how living things and their activities change with the seasons. Even though you may "think" you know what the world around you looks like, you don't actually know until you observe it. Also, the world around you changes daily, weekly, monthly, and yearly.

If you live in the city, buildings are constructed, torn down, and rebuilt. Weather fades colors that were once brightly painted. Trees grow and fall down. Grass grows and is mowed. In the country, fields are plowed and crops grow and are harvested. Animals have babies that grow to be adults. Storms create rivers, and snow and ice melt, sometimes creating floods.

By observing the world around you, you can learn about how geology works. The world becomes more interesting when you pay attention to what things look like, how they change, and how they stay the same.

V. Just For Fun

Imagine you got an email from someone on the planet Alpha Centauri Bb. They have never been to Earth, and they ask you what Earth is like. Using the data you have collected, write a narrative (story) describing what the area is like where you live. Include enough detail that the Alpha Centaurian can form a mental image of your surroundings.

Email to an Alpha Centaurian

Experiment 2

Hidden Treasure

Introduction

Make your own treasure map!

I. Think About It

❶ If you are hiking in a wilderness area where a GPS device doesn't work, what tools could you use to find your way? Why?

____

____

____

____

____

❷ What facts do you think you can learn about an area by using a map?

____

____

____

____

____

❸ If you were making a map of the area where you live, what details would you include? Why? How do you think someone else might use this map?

____

____

____

____

❹ What do you think you could discover if you had a topographic map, a compass, a rock hammer, and a rock and mineral test kit in your backpack?

❺ What advantages do you think you might have if you added a GPS device to your backpack? Why?

❻ What other electronic tools do you think you might use to help you explore the geology of an area? What do you think you might discover by using them?

II. Experiment 2: Hidden Treasure

Date ___________

Objective

____

____

Hypothesis

____

____

____

Materials

pencil, pen, colored pencils
compass
small jar or container with a lid
small items to place in the jar (treasure)
garden trowel (if needed)

EXPERIMENT

❶ Find some small objects to be your treasure and put them in the jar.

❷ Select an area near your home to make a map of. This area can be your front or back yard, a park, or other open space.

❸ Using each of your feet as a one-foot ruler, measure the outline of the area, walking heel-to-toe around it. Count your steps and notice if your path goes in a straight line or curves around objects.

After measuring a side of the area, draw that side in the space provided on the next page. Make your map as accurate as possible, noting on your map now many steps (feet) there are on each side of the area.

❹ Once you have finished measuring and drawing the outline of the area, hold the compass and turn around until the needle lines up with the north (N) symbol. Note this direction (north) on your map, drawing an arrow and an "N." Next, note south (S), east (E), and west (W), again drawing arrows and using the letters.

Treasure Map

❺ Add details to your map. Measure the distance to trees, shrubs, buildings, or other features by walking heel-to-toe and counting your steps. On the map, record the direction, the distance, and a drawing of the object. If you need to find out where to locate an object, such as a tree, on your map, pick a starting point (for example, the object in the corner of your map) and measure the distance as you walk toward the tree you want to position on the map.

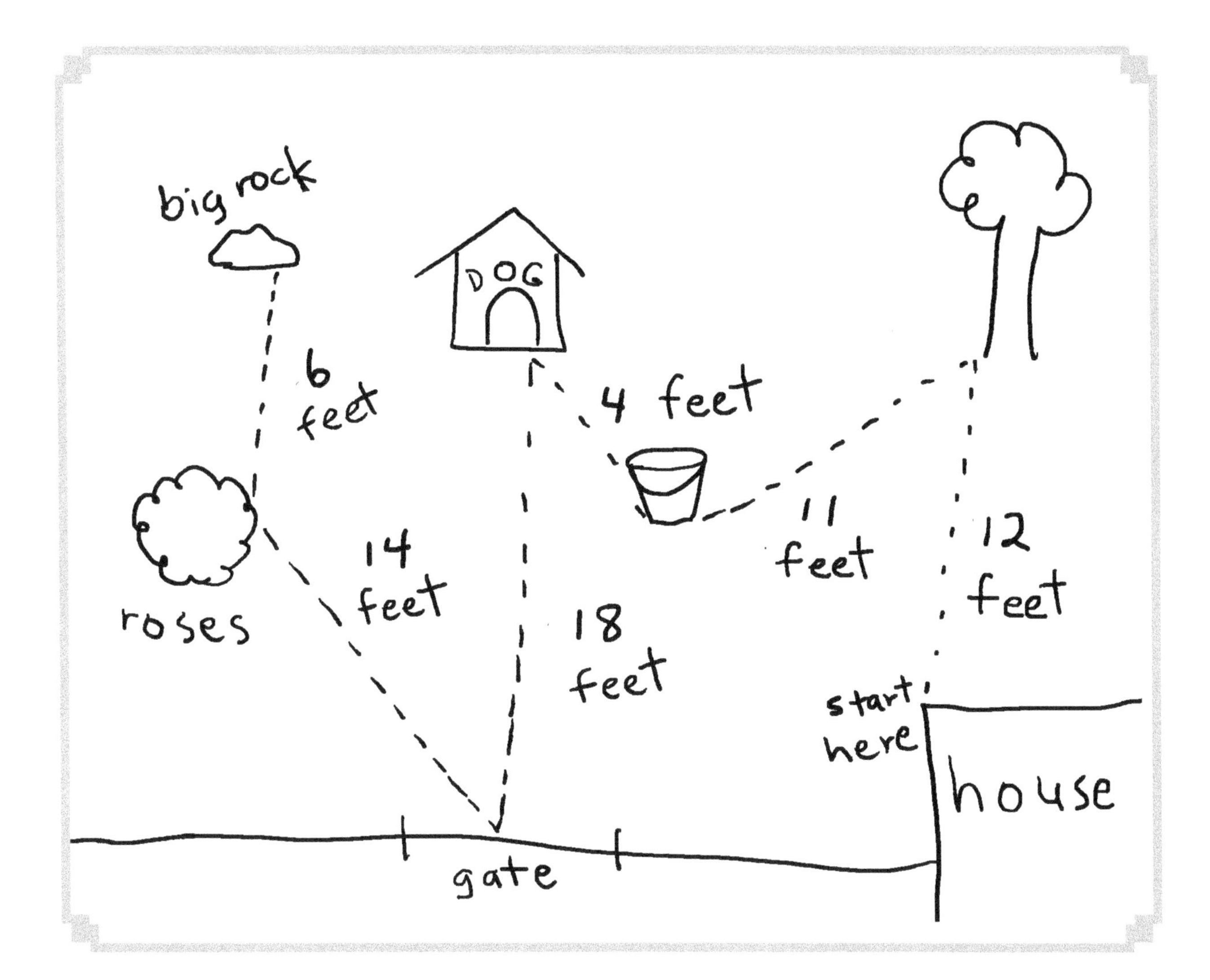

❻ Now pick a location in the area you have mapped and bury or hide your treasure there. Record the location of the treasure on your map.

❼ Give your map to a friend and see if they can use the map to find your hidden treasure.

Results

❶ Observe how easy or difficult it is for your friend to find your hidden treasure.

❷ In the following chart, record the number of attempts it takes your friend to find the hidden treasure. Also record how much help you have to give them before they find the treasure. Note any adjustments you make to your map to make it more accurate.

Treasure Map Results

Number of Attempts	
Type of Help	
Adjustments to Map	

III. Conclusions

What conclusions can you draw from your observations?

IV. Why?

As with any science, geology has been transformed as a result of technological advances. Modern tools allow geologists to study aspects of Earth's features and dynamics that were unavailable to early geologists.

Hand tools are easy to carry and easy to use, even if they are not necessarily technologically advanced. The rock hammer, map, compass, and rock and mineral test kit are tools that fit nicely in a backpack, are generally inexpensive, and are easy to use.

Modern geologists use a variety of electronic tools to study Earth. Although electronic tools are dependent on a variety of other technologies working properly (e.g., batteries, satellites, motors), they can offer substantial advantages over older or non-technological tools. For example, a global positioning system may find your location more accurately than you can find it on a paper map, and paper maps can become out of date. Some GPS devices can record the route you have walked and make it easier for you to find your way back than it would be if you were using a compass and map. However, GPS devices don't work everywhere and batteries may fail, so a geologist out in the field may also do well to know how to use a paper map and a compass.

Today's geologists also have ground penetrating radar available that can image below the surface of the Earth without disrupting the ground. GPR can be used from satellites as well as from the surface of the Earth. Satellite GPR and other remote sensing devices can help geologists find out about things such as the below surface structure of landforms and where to find groundwater, even in locations that are difficult to access. Archaeologists are now using satellite GPR to find ancient ruins that have been buried over the years.

Other tools used by geologists include drills and rock and mineral test kits. These tools help geologists directly test samples taken from the Earth to find out more about the composition and structure of landforms in different areas. Seismometers and seismographs help geologists understand what makes earthquakes and volcanoes occur and what the layers of the Earth may be like.

As technology advances, geologists will be able to find out more and more about Earth's structure and dynamics.

V. Just For Fun

Review the results of the experiment and evaluate your map. When your friend looked for the treasure, what parts of the map worked and what parts didn't work? Think of as many ways as you can to make your map better. Indicate a new location to hide the treasure. Now have a friend use the map to find the treasure in its new location.

Evaluate your revised map and record your results.

Treasure Map Results #2	
Number of Attempts	
Type of Help	
How well did the revised map work compared to the original map? Why?	

Experiment 3

Mineral Properties

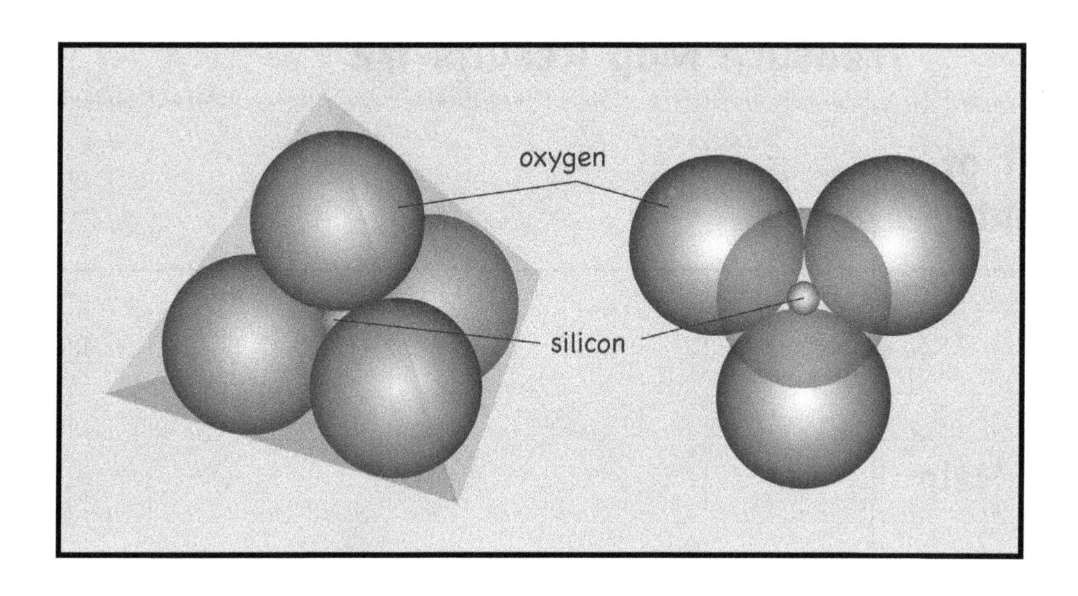

Introduction

Explore using scientific tests to identify minerals.

I. Think About It

❶ What do you think dirt (soil) is made of? Why?

❷ Do you think soil everywhere on Earth is the same? Why or why not?

❸ What do you think rocks are made of? Why?

❹ Do you think if you went outside with a shovel, you could dig up some minerals? Why or why not?

❺ Do you think minerals can be used to make artworks? Why or why not?

❻ Do you think there are ways to tell one mineral from another? Why or why not?

II. Experiment 3: Mineral Properties

Date ___________

Objective

____

____

Hypothesis

____

____

____

Materials

known mineral samples:
calcite, feldspar, quartz, hematite
several rocks from your backyard or near your home
copper penny
steel nail
streak plate (unglazed white ceramic tile)
paper
scissors
marking pen
tape
vinegar
lemon juice
eyedropper or spoon

EXPERIMENT—Part I

- The hardness of a mineral is determined by its resistance to being scratched. The Mohs scale of mineral hardness lists the relative hardness of ten common minerals and also various objects that can be used in scratch testing minerals.

  Test the hardness of the known mineral samples shown in the materials list by using the objects listed in the Mohs hardness scale on the next page. Fill in the chart on the next page with your test results and the appearance of the sample (color, texture, size, etc.).

Results

Mohs Scale of Mineral Hardness	
Object	**Hardness**
fingernail	2.5
copper penny	3
steel nail	5.5
streak plate	6.5

Mineral Name	Hardness	Appearance
Calcite		
Quartz		
Feldspar		
Hematite		

EXPERIMENT—Part II

- Do a streak test for each of the known minerals. Take each mineral sample and rub it firmly across the streak plate.

Results

Record the streak color of each mineral sample.

Mineral Name	Streak Color	Notes
Calcite		
Quartz		
Feldspar		
Hematite		

EXPERIMENT—Part III

❶ Go to the backyard or other place nearby and collect several rocks that look different from each other in color and texture.

❷ Using paper and tape or some other method, label the rocks you collected. Use a different number to identify each rock. Record the numbers in the following chart.

❸ Using the Mohs scale of mineral hardness, do a scratch test to find out the hardness of each unknown sample. Next do a streak test.

❹ Record your results. Review your results from Part I and Part II and see if you can tell what mineral is in your rock sample.

Results

Rock #	Hardness	Streak Color	Possible Mineral

III. Conclusions

In this experiment you tested the hardness and streak color of known and unknown minerals. By comparing your results from Parts 1-III of the experiment, were you able to determine the type of mineral or minerals that are in the rocks you collected near your home? Why or why not? Record your conclusions.

The hardness test using the Mohs hardness scale and the streak test are "subjective" tests. That is, the outcome may vary depending on the type of ceramic tile used for the streak plate, the type of nail or penny, and individual interpretations of results.

Another way to test for the type of mineral is to use chemical analysis. An acid test uses hydrochloric acid to test for calcium carbonate. Minerals containing calcium carbonate will effervesce (bubble or foam) when acid is used to create a chemical reaction. Do you think this is a more objective type of test? Why or why not? Write your answers below.

IV. Why?

In the same way that different plastics are made of different molecules that make them harder or softer and different colors, minerals made of different atoms and molecules will be harder or softer and different colors. The Mohs scale of mineral hardness is a quick and easy test used by geologists to help determine the type of minerals in a sample. The streak color and the acid test are also quick ways to help with mineral identification. The materials used to perform these tests are easily carried in a backpack out in the field.

By using simple tests and observations, field geologists can quickly determine if the rocks in an area are most likely limestone, granite, mica, or feldspar. Samples can be taken back to the lab for a more complete analysis requiring more complicated equipment.

V. Just For Fun

Use vinegar to test your rock and mineral collection to see if any samples have a chemical reaction with the acid. Use an eyedropper or spoon to apply the liquid. Record your results in the following chart.

Now use lemon juice to test your mineral collection for a chemical reaction. Record your results.

Compare your results. Did your samples react the same way to the vinegar as they did to the lemon juice? Why?

Look online or at the library to find out how your known minerals would be expected to react with acids. Did your known minerals react in the expected way? Why or why not?

Acid Test

Mineral	Vinegar Reaction?	Lemon Reaction?	Expected Reaction & Notes
Calcite			
Quartz			
Feldspar			
Hematite			
Rock # ____			
Rock # ____			
Rock # ____			
Rock # ____			
Rock # ____			

Experiment 4

Model Earth

Introduction

Explore Earth's layers by building a model.

I. Think About It

❶ Do you think scientists will ever be able to get actual samples from all the different layers of the Earth? Why or why not?

❷ What do you think it would be like if the Earth were solid rock all the way to the core and there were no different layers?

❸ Do you think by studying rocks geologists can learn anything about the interior of Earth? Why or why not?

❹ Do you think there are living things in the asthenosphere? Why or why not?

____

____

____

____

____

❺ What do you think Earth would be like if there were no crust or lithosphere?

____

____

____

____

____

❻ How do you think advances in technology have allowed us to discover more about the structure of Earth?

____

____

____

____

____

II. Experiment 4: Model Earth

Date ___________

Objective ____

____

Hypothesis ____

____

____

Materials

____

____

____

____

____

EXPERIMENT

In this experiment you will decide how to build an accurate model of Earth. Scientists use models to help them understand how things work. Creating accurate models is an important skill when doing science. The more accurately a model depicts reality, the more scientists can learn about how things work.

❶ On the following page, list what you know about Earth's layers. Record features such as whether scientists believe the layer is soft or rigid, solid or liquid, rock or iron. Also record the depth of the layer and any other features the layer may exhibit.

Use the *Student Textbook*, internet, and/or library to collect your information.

Features of Earth's Layers

Layer	Depth	Features
Crust		
Lithosphere		
Asthenosphere		
Mesosphere		
Inner Core		
Outer Core		

❷ Using the chart that you created on the previous page, go through the information you collected about each layer. Think about what material you could use to accurately represent each layer. Record the materials in the chart below.

Model Materials

Layer	Materials
Crust	
Lithosphere	
Asthenosphere	
Mesosphere	
Inner Core	
Outer Core	

❸ Decide which layers of Earth you will represent in your model. In the *Materials* section on the first page of this experiment, list the materials you will use for your model.

❹ Assemble a model of Earth using these materials.

Results

1. Review your model and observe how well it represents Earth's geology.

2. In the chart below write your observations about how accurately your model depicts the overall architecture of Earth and the characteristics of each layer.

Model Results

Questions	Observations
Does your model have layers? Which ones? Describe each layer.	
Do the layers represent accurate depths in your model? How do you know?	
Do the layers represent accurate consistency? (For example, is the lithosphere rigid and the mesosphere soft?)	
Do the layers in the model accurately represent the change that occurs at the boundary between each of the layers? Why or why not?	

III. Conclusions

Based on your observations, evaluate the accuracy of your model. How easy or difficult was it to accurately represent Earth's different layers?

IV. Why?

In this chapter you explored how to build an accurate model of Earth. As you know, model building is an important part of doing science. Models help scientists understand how things work. However, it is not always easy to build accurate models. For example, Earth is much too large to build a model of the exact same size as Earth. Also, scientists don't know for sure what layers below the crust look like, so building an accurate model of Earth is difficult. Parts of the model will be accurate and other parts will most likely be inaccurate.

As scientists learn more and more about Earth's layers, better and more accurate models will be developed.

V. Just For Fun

Make another model of Earth, but this time make one you can eat!

CHOCOLATE LAVA CAKE

butter 113 grams (1/2 cup)
semi-sweet chocolate chips 133 ml (1/2 cup + 1 Tbsp.)
2 whole eggs
2 egg yolks
powdered sugar 192 ml (3/4 cup + 1 Tbsp.)
flour 94 ml (1/3 cup + 1 Tbsp.)

Microwave butter briefly until melted. Stir in chocolate chips until melted. Mix in whole eggs and yolks, then powdered sugar. Stir in flour. Pour into custard cups thoroughly greased with butter. Bake at 190° C (375° F) until edges are set and centers are still soft, about 10-13 minutes. Don't overbake.
Makes about 4.

Can you think of any food items you can add to represent the Earth's crust and inner core?

On the next page record the accuracy of this model and then compare it your first model.

Model Results and Comparisons

Experiment 5

Dynamic Earth

Introduction

Explore plate tectonics with this experiment.

I. Think About It

❶ What do you think Earth would be like if the lithosphere were one solid piece instead of being broken into plates?

❷ What do you think would happen if there were no convection in the soft layers of Earth?

❸ What do you think would happen if all lava were the same viscosity?

❹ What do you think Earth would be like if there were no earthquakes and volcanoes?

❺ What do you think you might learn by studying the ground after an earthquake?

❻ What do you think you might learn by observing a volcano erupting?

II. Experiment 5: Dynamic Earth

Date ___________

Objective

____

____

Hypothesis

____

____

____

Materials

brittle candy (recipe below)
1 jar smooth peanut butter
118 ml (1/2 cup) crushed graham crackers

EXPERIMENT

❶ Use the following instructions to make brittle candy.

Brittle Candy Recipe

Ingredients

237 ml (1 cup) white sugar
118 ml (1/2 cup) light corn syrup
1.25 ml (1/4 teaspoon) salt
59 ml (1/4 cup) water
28 grams (2 Tbsp) butter, softened
5 ml (1 teaspoon) baking soda

Equipment

2 liter (2 qt) saucepan
candy thermometer
cookie sheet, approx. 30x36 cm (12x14 inches)
2 spatulas

Brittle Candy Recipe—Instructions

Grease a large cookie sheet.

Measure sugar, corn syrup, salt, and water into a heavy 2 liter (2 quart) saucepan. Bring mixture to a boil over medium heat.

Stir until the sugar is dissolved.

Put the candy thermometer in the saucepan and continue cooking.

Stir frequently. The candy should be done when the temperature reaches 150° C (300° F). You can check whether it is done by dropping a small amount of the hot candy mixture into very cold water. If the candy separates into hard and brittle threads, it is done cooking.

Remove from heat and quickly stir in butter and baking soda.

Pour at once onto the greased cookie sheet and spread the mixture into a rectangle of about 30x36 centimeters (12x14 inches). You can use two buttered spatulas to spread the candy.

Allow to cool. Break the candy into large pieces.

❷ Spread a 1.25 cm (1/2 inch) thick layer of peanut butter on a plate or another cookie sheet.

❸ Mix 118 ml (1/2 cup) of crushed graham crackers with 59 ml (1/4 cup) of peanut butter.

❹ Take 2 large pieces of brittle candy and spread the peanut butter/graham cracker mixture on top of each of them.

❺ Place the two pieces of brittle candy (peanut butter/graham cracker side up) about 2.5-5 cm (1-2 inches) apart on top of the peanut butter that is on the plate or cookie sheet.

❻ In the following chart write which layer each food item represents (the asthenosphere, the lithosphere, or the crust).

Earth's Layers Represented

Item	Earth's Layer
Peanut Butter	
Brittle Candy	
Graham Cracker/Peanut Butter Mixture	

❼ Gently holding the sides of the brittle candy pieces, move the candy around on top of the peanut butter and observe what happens.

Have the pieces bump into each other, scrape alongside each other, and move up or down with respect to each other.

Move the pieces quickly and slowly and observe the difference. Try to get one piece to slide under the other piece and observe what happens to the graham cracker topping.

❽ In the box on the next page, write about what you observe during this experiment. Think about what you have learned about plate tectonics and how your model relates to what you know.

Plate Tectonics Model—Observations

Results

Review your observations and answer the questions in the following chart. Note what happened to the lower peanut butter layer, the brittle candy pieces, and the top graham cracker/peanut butter layer.

Tectonic Plates Model Results

Questions	Observations
What happened when you moved the brittle candy pieces slowly on top of the peanut butter?	
What happened when you moved the brittle candy pieces quickly on top of the peanut butter?	
What happened when you collided two brittle candy pieces together?	
What happened when you moved the pieces up or down with respect to each other?	
What happened when you slid the two pieces side by side in different directions?	

III. Conclusions

What conclusions can you draw from your observations?

Based on your observations and results, explain what you think happens when two of Earth's plates collide with each other, slide past each other, or move up and down with respect to each other. Do you think your model was a good representation of plate tectonics? Why or why not?

IV. Why?

Earth is a dynamic planet, meaning it is constantly changing. Features of Earth's surface can be changed quickly by natural phenomena, such as sudden earthquakes and volcanic activity. There are also changes that occur over very long periods of time, such as the formation of river valleys, the wearing away of land forms by glaciers, and the growth of mountains due to forces caused by the movement of tectonic plates.

Plate tectonics is a theory that has been proposed to explain how earthquakes, mountain ridges, and volcanoes occur. Although scientists can't directly sample the layers below the Earth's crust, by putting together pieces of information they obtain by making observations, they can propose a theory to explain dynamic phenomena. Scientists can draw conclusions based on data from observations such as the location, age, and activity of volcanoes; analysis of the composition and deformation of rock in mountains; and the location, type, and strength of earthquakes.

According to the theory of plate tectonics, earthquakes occur when two plates collide with, push on, or slide past each other, causing stresses to build up along the plate boundaries and also in the interior of the plates. Earthquakes occur along faults, or fractures, in the Earth's surface. The most active faults are at or near plate boundaries, such as the San Andreas Fault in California. But there are also some active areas within plate interiors, such as the New Madrid Seismic Zone located in the Mississippi Valley.

The movement of plates can cause mountains to form as land masses are pushed together, causing the land to fold. Mountains can also be formed as the movement of plates causes big blocks of land to move up and down.

Although the model in this experiment is not a perfect representation, it is helpful in visualizing how the layers and forces within the Earth work to create earthquakes and form mountains.

V. Just For Fun

Using the internet or the library, research various ways to create a model volcano. Choose a method and write your own experiment.

Experiment 5: Model Volcano

Date ___________

Objective ______________________________

____

Hypothesis ______________________________

____

____

Materials

____

____

____

____

EXPERIMENT

____

____

____

____

____

____

EXPERIMENT (continued)

Model Volcano

III. Conclusions

Experiment 6

Using Satellite Images

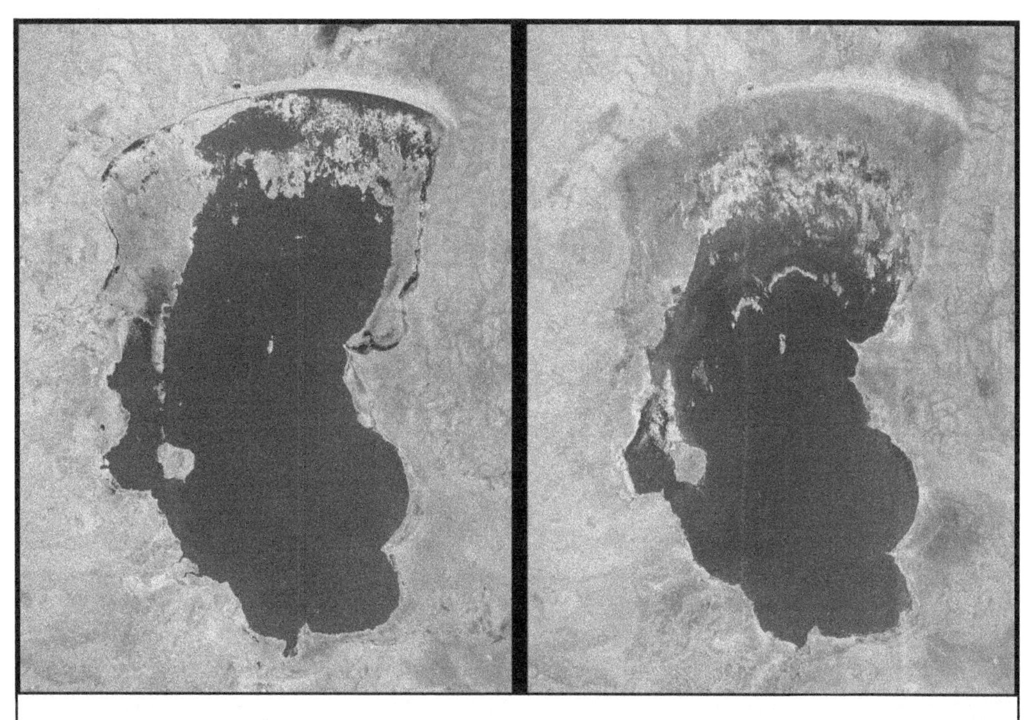

Two satellite images of Lake Chilwa, Malawi show it is shrinking
USGS Landsat photos

Introduction

Explore using satellite imagery to study Earth's spheres.

I. Think About It

❶ List a few events that you think could change the geosphere.

❷ List a few events that you think could change the atmosphere.

❸ List a few events that you think could change the biosphere.

❹ List a few events that you think could change the hydrosphere.

❺ Do you think changes to the geosphere can occur below the Earth's crust? Why or why not?

❻ Do you think there are events that happen in space that can affect any of Earth's spheres? Why or why not?

II. Experiment 6: Using Satellite Images

Date ___________

Objective

____

____

Materials

computer with internet access
printer and paper (optional)
colored pencils (optional)

EXPERIMENT

The US Geological Survey (USGS) operates a satellite called Landsat that photographs Earth's surface. Their online Land Remote Sensing Image Gallery shows sets of images that compare changes to Earth over time.

❶ Go to the US Geological Survey Landsat website:

https://remotesensing.usgs.gov/gallery/

Spend some time looking through the collections of images that show changes to Earth's surface.

❷ Select one set of images that shows a change to the geosphere. In the *Download Image* section, click on the small size image file and download it. If you get a "zip file" that contains the images, you will need a program that unzips files for you to be able to look at the images. Make a folder on your computer to keep the image files in.

If possible, print the images, label them, and insert them in your *Laboratory Notebook*. If you can't print them, you can refer to the files on your computer as needed.

Record your observations in the *Results* section.

❸ Select one set of images that shows a change to the atmosphere. Download and print them if possible. Record your observations in the *Results* section.

❹ Select one set of images that shows a change to the hydrosphere. Download and print them if possible. Record your observations in the *Results* section.

❺ Select one set of images that shows a change to the biosphere. Download and print them if possible. Record your observations in the *Results* section.

Results

A Change in the Geosphere

Location:

Period of time covered:

Description of changes to the geosphere. Include a rough sketch.

How do you think this information could be used to help the environment and/or to help people?

A Change in the Atmosphere

Location:

Period of time covered:

Description of changes to the atmosphere. Include a rough sketch.

How do you think this information could be used to help the environment and/or to help people?

A Change in the Hydrosphere

Location:

Period of time covered:

Description of changes to the hydrosphere. Include a rough sketch.

How do you think this information could be used to help the environment and/or to help people?

A Change in the Biosphere

Location:

Period of time covered:

Description of changes to the biosphere. Include a rough sketch.

How do you think this information could be used to help the environment and/or to help people?

III. Conclusions

Discuss what you learned by observing images taken of Earth from space. How do you think satellites have changed what we can learn about Earth?

IV. Why?

One way scientists observe changes in Earth's spheres is through satellite images. By using images from space, scientists can watch natural events such as hurricanes, typhoons, dust storms, volcanoes, and wildfires and how they affect the different spheres of Earth. Changes in forests, deserts, glaciers, rivers, icebergs, oceans, pollution, and population can be observed over time to learn how natural and human activity affect the atmosphere, hydrosphere, biosphere, geosphere, and magnetosphere.

By examining images taken of the same location over the course of several days, weeks, months, or years, scientists studying different parts of Earth's spheres can learn how changes to Earth's interconnected system occur over time. For example, if a particular typhoon were followed for several days, observations could be made about where and how it forms and grows, what route it follows, and where and when it breaks apart. This information could help scientists predict when future storms might occur. The effects of the passage of the typhoon could also be studied, showing how the biosphere, geosphere, hydrosphere, and atmosphere were affected.

Using satellite images to study different natural events allows scientists to gather information that can be used to help people be better prepared for storms and other natural events and help them find ways to protect their communities and surrounding environments. The effects of population growth can be seen as cities and towns expand, and satellite images may be used to help find solutions to the impact on the environment. Air and water pollution can be observed, helping solutions to be found, and the effects of deforestation and other changes in plant distribution can be studied.

In addition to photographs, satellites use an array of other technologies, such as radar, laser beams, and thermal imaging, to further record details of Earth's system. Putting together information from these different types of imagery gives scientists a more complete picture of Earth and its spheres.

V. Just For Fun

Discover more about Earth's spheres through satellite images.

NASA's Earth Observatory website has photographs and videos taken from the International Space Station.

http://earthobservatory.nasa.gov/

On the top menu bar click on *Images* to find collections by topic. Select *Natural Hazards* and explore satellite images of several natural disasters.

Go back to the Home pages and scroll down to find the *Special Collections* groups of photographs to explore.

NASA's Gateway to Astronaut Photography of Earth

http://eol.jsc.nasa.gov/Collections/EarthFromSpace/

Select a topic from the left menu bar. On the next screen select what you would like to view and click *Start Search*. Play around to see what else you can find on this site. Record some of your observations.

Observations of Earth from Space

What did you see that was the most surprising?

Observations of Earth from Space

What did you see that was the most beautiful?

What would you like to explore further?

What could you notice that you wouldn't be able to see from the ground?

If you were a geologist, how do you think you would use some of the satellite images you saw?

Experiment 7

Modeling Shaky Ground

A building collapsed on a car during an earthquake, Loma Prieta, CA

Courtesy of USGS by J. K. Nakata

Introduction

Learn more about the geosphere by making a model.

I. Think About It

❶ Have you ever experienced an earthquake? If so, what was it like? If not, what do you think it would be like?

❷ Do you think the way buildings and roads are designed can help minimize damage from an earthquake? Why or why not?

❸ Do you think scientists can predict where an earthquake might happen? Why or why not?

❹ Do you think scientists can predict when an earthquake might happen? Why or why not?

❺ What difficulties do you think scientists face when studying earthquakes?

❻ Do you think science will someday be used to prevent earthquakes? Why or why not?

II. Experiment 7: Modeling Shaky Ground

Date ___________

Objective

____

____

Hypothesis

____

____

____

Materials

Jell-O or other gelatin (any color/any flavor)
graham crackers
marshmallows (Optional: large and small)
toothpicks
baking pan — 24 cm x 28 cm (9.5" x 11")

EXPERIMENT

❶ Mix the gelatin according to the instructions on the box. Pour the liquid gelatin into the baking pan and refrigerate until firm.

❷ Remove the gelatin from the refrigerator. Place graham crackers on top of the gelatin. Observe how they are arranged. (Are they touching? Far apart? Flat? At an angle? What else can you notice?) In the box below, draw and label your observations.

Observations—Jell-O and crackers

Example of basic diagram of experimental setup:

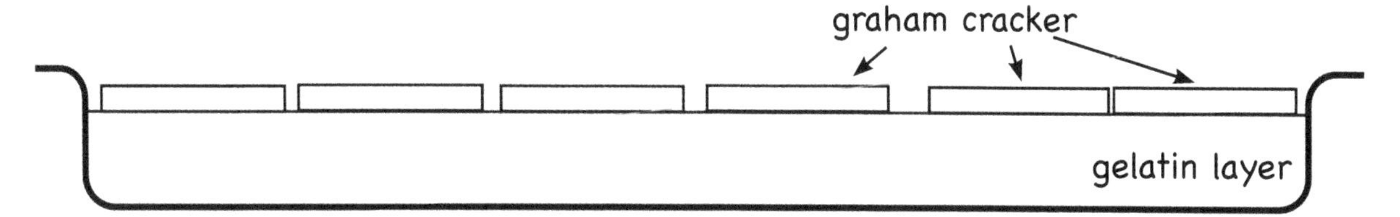

❸ Press down on the graham crackers at one edge of the pan and observe what happens. (How far down does the graham cracker move? When you press on one graham cracker, do the others move? How? What else can you observe?)

❹ Repeat several times, experimenting with pressing on different crackers, more than one at a time, using more or less pressure, pulsing the crackers, etc.

❺ In the following two boxes, draw and label your observations. Use additional sheets of paper for more observations.

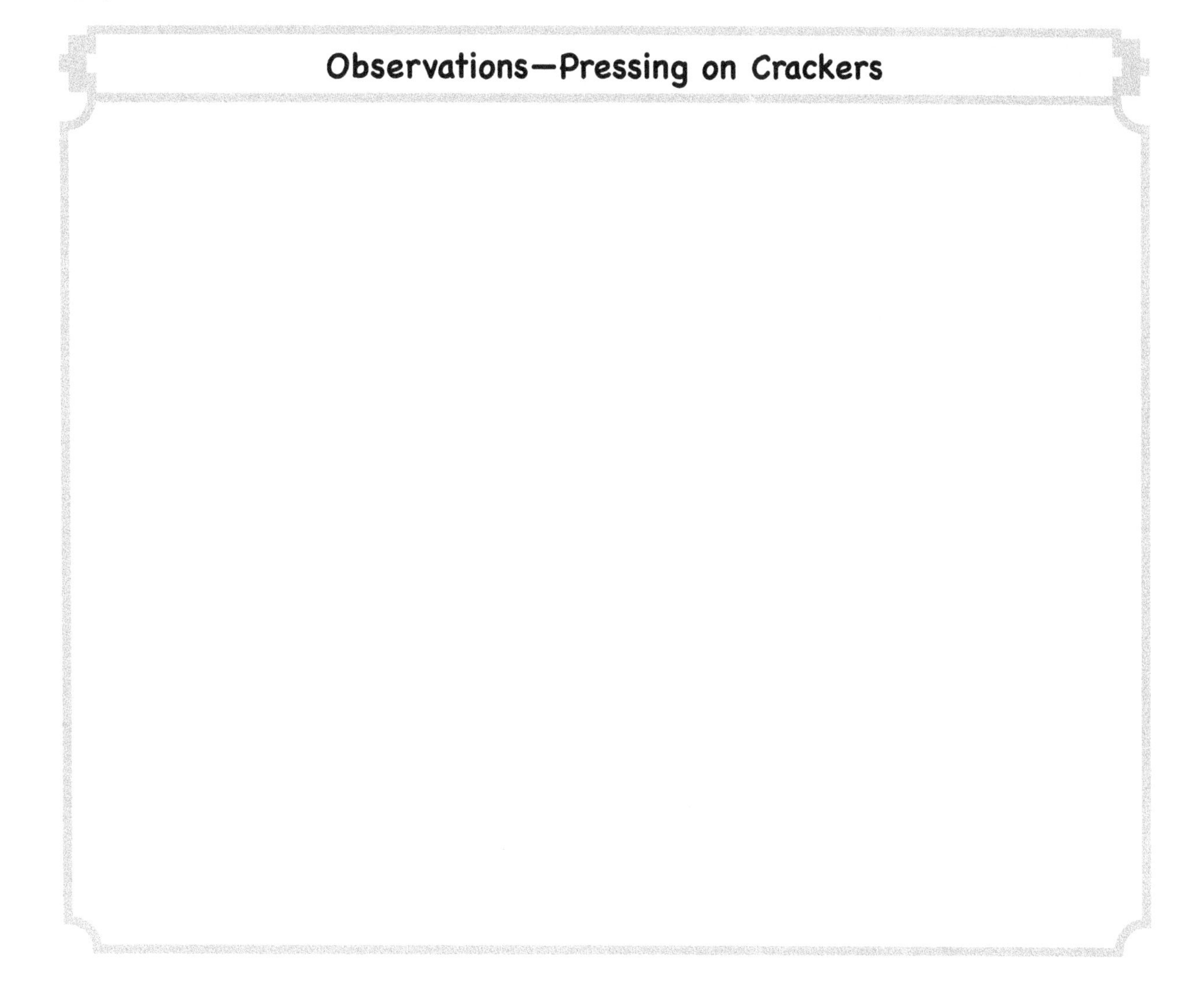

More Observations—Pressing on Crackers

❻ Place marshmallows on the graham crackers. Experiment with pressing on different crackers, more than one at a time, using more or less pressure, pulsating the pressure, etc. What happens to the marshmallows?

Record your results in the *Results* section.

❼ Try stacking the marshmallows and repeat Step ❻. What happens?

Record your results in the *Results* section.

❽ Using toothpicks, secure the marshmallows to the graham crackers. Repeat Step ❻. What happens? Can you get the marshmallows to move?

Record your results in the *Results* section.

Results

Observations—Marshmallows

(Continue on next page.)

More Observations—Marshmallows

(Continue on separate sheets of paper.)

III. Conclusions

Discuss what you learned by creating a model of Earth. What does the gelatin represent? What do the graham crackers represent? What does this model represent? What are some limitations of this model and models in general?

IV. Why?

Model building is an important aspect of scientific investigation. Scientists build models of their ideas to help them see how things work and how to think of new ways to understand scientific phenomena. Scientists can use models to understand how Earth's layers may interact to create earthquakes.

However, model building isn't always easy, and finding the right materials to model important features can be a challenge. For example, you might have elected to use modeling clay instead of gelatin to model the soft layer of the asthenosphere and then found it difficult to create movements in the modeling clay that would create earthquake-like movements of the graham cracker plates. Or you might have decided to make plates of modeling clay and found them to not be rigid enough for your experiment.

Models don't generally duplicate all the important features of the idea or object a scientist is trying to understand. Knowing which features to include and which to ignore depends on what the model is being used to explore.

V. Just For Fun

Think about how you might change this experiment to learn more about earthquakes. For example, what would happen if, instead of using marshmallows, you built different types of structures from different materials? Could you use something besides graham crackers for the plates? What would happen if you made gelatin of different thicknesses by adding more or less water when you mixed it? What else can you change?

Create your own experiment. Use separate sheets of paper, give your experiment a name, and follow the outline of the Modeling Shaky Ground experiment.

Experiment 8

Exploring Cloud Formation

Courtesy of US Fish and Wildlife Service, photo by Steve Hillebrand

Introduction

Do you think you can make clouds in a bottle?

I. Think About It

❶ How do you think clouds are formed?

❷ How many different kinds of clouds have you observed? How would you describe them?

❸ What factors in the atmosphere do you think affect cloud formation?

❹ Do you think clouds are more likely to form over an ocean or a desert? Why?

❺ What do you think would happen to life on Earth if there were no clouds? Why?

❻ How do you think clouds move? Do you think they go far? Do you think they go fast? Why or why not?

II. Experiment 8: Exploring Cloud Formation

Date ___________

Objective

____

____

Hypothesis

____

____

____

Materials

2 liter (2 quart) plastic bottle with cap
warm water
matches

EXPERIMENT

❶ Pour warm water into the plastic bottle until it is about 1/4 full. Put the cap on the bottle.

❷ Light a match and remove the cap from the bottle. Drop the match in the bottle and quickly replace the cap.

❸ Squeeze the plastic bottle near the bottom and release. Notice what happens to the air in the bottle as you do this.

❹ Record your observations in the chart in the *Results* section.

❺ Repeat this experiment, filling the plastic bottle 1/2 full, 2/3 full, and then almost full. After each experiment, empty the bottle and start with fresh warm water. Record your observations each time.

Results

Cloud Formation Observations

Water Level in Bottle	Observations
1/4 Full	
1/2 Full	
2/3 Full	
Almost Full	

III. Conclusions

What conclusions can you draw from your observations? How would you relate your results to what happens in the atmosphere?

IV. Why?

Water vapor is essential for life. It is the most important gas for keeping Earth warm and is also part of Earth's water cycle. Liquid water evaporates from bodies of water and from the land and enters the atmosphere as water vapor. The water vapor later condenses, turning back to the liquid state, forming clouds and then falling to Earth as precipitation.

Since warm air can hold more water vapor than cold air, cooling the air to the point where it can no longer hold all the water vapor will cause the water vapor to condense and turn to the liquid state. The temperature at which condensation occurs is called the dew point. The dew point varies depending on the amount of water vapor in the air (humidity).

The most common way for air to be cooled is through lifting. Air moves into an area of the atmosphere that is at a lower pressure, causing the air to expand. Energy is required for this expansion, taking heat away from the air and cooling it. As the air cools, some of the water vapor will condense around dust particles in the air, forming water droplets.

The reverse happens as air sinks. As it encounters higher pressures at lower altitudes, the increased pressure squeezes the air, adding heat and allowing the air to once again hold more water vapor. This can cause clouds to evaporate as the liquid water turns to vapor.

In order for water droplets to fall as precipitation, they need to become larger and heavier. One process by which this happens is called the collision and coalescence process or the warm rain process. During this process, water droplets in clouds collide and stick together (coalesce) to form larger drops.

Since warm air can hold more water vapor than cold air and heat makes water evaporate, the highest levels of water vapor in the atmosphere are over the oceans in the equatorial region where the heat evaporates water from the oceans and more of this water vapor can be held in the warm air. The lowest levels of water vapor are found over the dry deserts where there is little water to evaporate and at the poles where the air is cold and the water is tied up as ice.

V. Just For Fun

Observe the clouds in your area for two weeks or more and record your daily weather observations. Describe the weather conditions and describe and sketch the clouds that you see. Follow the format below to continue your chart on separate paper.

Check online daily for humidity, dew point, and low and high temperatures. To find this information you can do a simple search such as "dew point today" with your location. Record this information on your chart.

At the end of your observation period, look at the data on your chart. Is there a relationship between the types of clouds and the weather? Do you think the types of clouds are related to the temperature, dew point, and/or humidity? Record your conclusions and fasten the observation chart pages in the *Laboratory Notebook*.

Cloud Formation Observations

	Observations (Weather, Types of Clouds)
Date: ________ Humidity: ______ Dew point: ______ High Temp: ______ Low Temp: ______	

Experiment 9

What Makes an Aquifer?

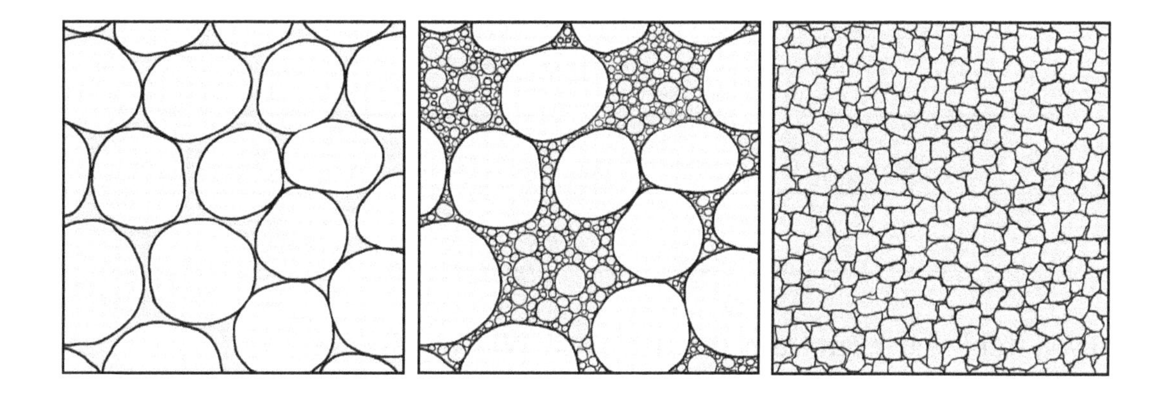

Introduction

See how water travels through some materials that make up Earth.

I. Think About It

❶ Do you think water can travel through soil? Why or why not?

❷ Do you think water can travel through rock? Why or why not?

❸ Where do you think the water you drink comes from? What steps are needed to get it to your house?

❹ What do you think would happen if there were no groundwater but only surface water? Why?

____

____

____

____

____

____

❺ If you lived where there was a shortage of water, what things could you do to use less and help the water supply last longer?

____

____

____

____

____

____

❻ Have you seen water pollution in your area? What actions do you think could be taken to help reduce water pollution and clean up polluted water?

____

____

____

____

____

____

II. Experiment 9: What Makes an Aquifer? Date ________

Objective ______________________________

____

Hypothesis ______________________________

____

Materials

gravel, about 600 ml (2.5 cups)
sand, about 600 ml (2.5 cups)
dirt (soil), about 600 ml (2.5 cups)
pottery clay, about 600 ml (2.5 cups)
water
(4) Styrofoam cups, about 355 ml (12 ounce) size
(4) 16 oz. clear plastic cups, glasses, or other clear containers
pencil
marking pen
measuring cups
graduated cylinder, 100 ml
large bowl

Optional

stopwatch or clock with a second hand

EXPERIMENT

Part I

This part of the experiment is a test for permeability of different materials.

❶ Use the pencil to poke several holes in the bottom of each of four Styrofoam cups. Put the same number of holes in each cup and make all the holes about the same size.

❷ Put 250 ml (1 cup) of each material (sand, gravel, dirt, and clay) in its own cup. Label each cup with the material it contains.

3. One at a time, hold each cup over a large bowl and pour 120 ml (4 ounces) of water into the cup. Note how long it takes the water to drain through each material. You can use a stopwatch or a timer with a second hand, or you can visually note how quickly or slowly the water runs through each material compared to the others.

4. In the *Results* section, note how long it takes the water to go through each material and write down your observations of the properties of the material.

Results

Permeability Test

Material	Relative Permeability & Properties

Part II

This part of the experiment is a test for porosity. Porosity describes the size and quantity of void, or empty, spaces between the particles of a material. The porosity of a material describes the quantity of liquid it can hold, and the permeability of the material determines the speed at which liquid can flow through the material. It is possible for a material to be porous but not permeable (impermeable) if the pores (void spaces) are not connected.

❶ Take four 16 oz. clear plastic cups, glasses, or other clear containers. Measure 350 ml (1.5 cup) of each material (sand, gravel, dirt, and clay) and put it in its own cup. Label each cup with the material it contains.

❷ Measure 100 ml of water in the graduated cylinder. Take one of the containers and slowly pour water into the material in it, trying to avoid making a depression in the top surface of the material. Pour in water until the water is up to the top of the material in the container. How much water is left in the graduated cylinder? Record your observations in the Results section.

❸ Repeat Step ❷ for the other three materials.

Results

Porosity Test

Material	Amount of water left in graduated cylinder

Use the following chart to calculate the porosity of each material.

❶ Calculate how much water was held in the pores of the first material you tested. To get this number, subtract the amount of water that was left in the graduated cylinder from the 100 ml of water you started with.

❷ Divide the amount of water you were able to pour into the material by the total amount of material. Express the answer as a percentage.

For example: If you were able to add 90 ml of water to one of the materials, the calculation would be:

90 ml water ÷ 350 ml material = .2571 = 25.71% porosity.

❸ Repeat Steps ❶-❷ for the remaining three materials.

Porosity

Amount of water held in pores	Porosity percentage

III. Conclusions

Which material was the most permeable? ______________________

Which material was the least permeable? ______________________

Which material was the most porous? ______________________

Which material was the least porous? ______________________

By doing this experiment, what did you learn about permeability and porosity? Do you think permeability and porosity are related? Why or why not? What other conclusions can you draw from your observations?

IV. Why?

In this experiment you probably discovered that of the materials you tested clay is the least porous and the least permeable, and gravel is the most porous and the most permeable. Clay is made of very small particles packed closely together. This leaves very little pore space to hold water. Also, because the space between particles is so small, it is difficult for water to travel through clay because it is less permeable.

Gravel contains large particles that form larger spaces between them. This allows more water to enter the gravel. Because the pore spaces are so large, water flows rapidly through gravel.

Soils are a mixture of different materials. Soils that are mostly sand will allow water to run through rapidly, making them dry quickly, and they will not have many nutrients to support plant life. In a soil that is made mostly of clay, water does not move well through it, and the tiny pores in the clay can make it difficult for plant roots to push through. The best soil for growing plants has the right mixture of materials that allow the soil to retain enough water for the plants to use, drains well enough that plants don't get too much water, and has enough organic matter to provide nutrients for the plants.

From this experiment you can see that the materials an aquifer is made of will affect the quantity of water the aquifer can hold and how fast water can flow through it. Also, the type of soil above the aquifer will affect how fast the aquifer can recharge. A soil that is mostly clay won't allow much precipitation to infiltrate and a gravely soil will allow water to flow through quickly.

V. Just For Fun

Build aquifers!

❶ Aquifers hold and carry water under the ground. Look at the results of your permeability and porosity tests and think about how you could layer the materials you tested to make an effective aquifer.

❷ Make a trough for building your aquifer. You can fold a piece of cardboard into a U shape and cover it with plastic. First decide how long, wide, and deep you want the aquifer to be and then have an adult help you cut the cardboard accordingly. Or you may have another idea about how you can build a trough.

❸ There are different types of aquifers, including those that hold the water in one area, those that let it flow short distances, and those that carry it long distances. The rate at which water flows also varies. Think about how you might use the materials you tested to create different types of aquifers.

❹ Use fresh samples of the materials for this part of the experiment. Choose two or more of the materials you tested and layer them in the trough. You may want to put screening or coarse cloth over the ends of the trough to keep the materials from washing out. This is your aquifer.

In the chart in the Results section, list the materials in the order in which they are layered in each aquifer.

❺ You may want to do this experiment outdoors or you can hold your aquifer at a slight angle with the lower end over a bucket. Pour water slowly on the higher end of your aquifer and note what happens. You can also try tilting the aquifer at a greater angle and pouring more water in the top end. What else could you try?

Record your observations in the *Results* section.

❻ Repeat Steps ❹-❺ to make several more aquifers. You can also add different materials of your choice to your aquifer.

Results

In the following charts record information about the layers in each of your aquifers and your observations about water flow, aquifer tilt, and any other variations.

When you have finished the experiment, review your data and write a summary of your observations. Then review your summary and based on these observations write conclusions about what you learned from making model aquifers. Also record how you think your model aquifers compare to real aquifers.

Aquifer

Aquifer Layers (Materials in order from top to bottom)	**Observations**

Aquifer

Aquifer Layers (Materials in order from top to bottom)	Observations

Aquifer

Aquifer Layers (In order from top to bottom)	**Observations**

Summary of Aquifer Observations

Conclusions

Experiment 10

My Biome

Introduction

Take a close look at the biome in which you live. What features make it your biome?

I. Think About It

❶ Do you think that you often pay attention to everything around you when you go outside? Why or why not?

❷ What do you think the area where you live would be like if there were no insects? Why?

❸ If the weather patterns during the course of several years became totally different from what they are now, do you think you would still be living in the same type of biome?

❹ Do you think animals behave in the same way every day of the year? Why or why not?

❺ Do you think there is a relationship between the kinds of animals and the kinds of plants found in your biome? Why or why not?

❻ How do you think the land formations and soils affect the plant and animal life in your biome?

II. Experiment 10: My Biome

Date ___________

Objective

____

____

Hypothesis

____

____

Materials

field notebook
pencil and colored pencils
small backpack
water bottle
snack

Optional

binoculars

EXPERIMENT

❶ Pack a small backpack with a water bottle, pencils, your field notebook, and a snack.

❷ Take a one to two hour hike in your surrounding environment.

❸ In your field notebook, record what you see. Pay attention to everything around you. Carefully observe the plant and animal life, where organisms live, and what they are doing. Observe the landforms, rocks, and soil and how these affect what lives in an area.

❹ Describe the types of plants in your surroundings and the types of animals (both large and small) and observe their interactions. Are ants crawling on plants? Are insects and animals eating plants? Are there dogs playing in the grass? Are cats looking for birds or insects to catch? Are birds pulling worms out of the ground? Are plants blooming?

❺ Observe the weather and how the weather affects the plant and animal life in your surroundings. Is there snow on the ground and do the plants look lifeless? Is it warm and sunny with lots of flowering plants? Are there insects out foraging for food? Are lizards sunning themselves? Is it raining? Or windy? Are the birds cheeping or sitting quietly waiting for a storm to pass? Record your observations in your field notebook.

❻ After your hike, review your notes and record your observations in the chart provided in the *Results* section.

Results

Summarize your observations below.

Summary of Biome Observations	
Route followed	
Weather	
Plants	
Animals	
Interactions	

Summary of Biome Observations

New things, interesting things, surprising things, and more!

III. Conclusions

Use the chart in the *Student Textbook* to determine the type of biome you live in. Record the name of your biome, use your observations to describe it, and list any unique or surprising features you discovered. Think about which features you would include if you were to describe your biome to someone who has never visited you.

IV. Why?

An environment is the set of conditions surrounding an organism in the region where it lives. An ecosystem is a specific area that contains a community of living things existing under similar conditions. An ecosystem can be any size—from very small to very large. On the other hand, a biome is always a very large region and is an ecosystem that is defined by the climate, soils, and plant life that exist within it.

By exploring your biome in this experiment, you were able to observe different plants and animals and the conditions under which they live. You probably noticed large animals such as dogs and people and very small animals such as insects; plants as large as trees and very small plants such as grass. You may have noticed what conditions different plants require to grow (how much sunlight, how much water, what type of soil) and whether some plants were being eaten by bugs or animals. You may have found a home that an animal built, such as a burrow or a bird's nest. With careful observation, you can begin to see how the different spheres of Earth interact to create the biome and how a change in certain conditions in the biome, such as the amount of rain, could affect the organisms that live there.

Think about what you observed on your hike and take another look at your notes. Did you describe different ecosystems within the biome? Did you pass a marsh, a pond, or other wet area? A meadow and a forest? Farmed land and front yards?

If we use the example of a temperate deciduous forest biome, we can see that although they are part of the same biome, a pond and a wooded area would be different ecosystems containing different communities of plant and animal life. Some plants and animals require the watery ecosystem of a pond, while others need an area that has a thick growth of trees. Within these different ecosystems different habitats will be found. A fish in the pond would be in a habitat with the right features to allow it to live; for example, the right kind of food, enough space, deep enough water (or perhaps shallow water), and a way to hide or escape from predators. The pond might have the right conditions for a water lily habitat and another habitat for a certain kind of frog to thrive. On the other hand, the wooded ecosystem would provide a place for very different types of habitats. Shade-loving plants could find a place to grow beneath the trees. Deer might live there, finding the plants they need to eat, places to sleep, and protection from predators. The trees could house many types of birds.

We can see that a biome is a very large area that includes many different ecosystems, with each ecosystem containing many different habitats. Figuring out the details of where an organism lives can get quite complicated!

V. Just For Fun

What's that bird?

Watching and identifying the birds around you is an interesting and fun way to learn more about ecosystems and habitats (and birds!). And you can bird-watch anywhere—even when you're riding in a car. Here are some suggestions for getting started.

- Get a field guide to the birds book that has good pictures and descriptions of the birds in your area. Many birders like the Peterson field guides, but other books are also good. Spend some time leafing through the book and familiarizing yourself with different types of birds. Which ones do you think you have seen in your area? When you go outside or look out your window, pay attention to the birds you see and then find them in the field guide. What can you learn about them? Binoculars of any kind are helpful when bird-watching.

- If you have a good location (and no outdoor cats), set up one or more bird feeders with different kinds of seeds and maybe a feeder for hummingbirds too. Tend the feeder daily and observe the birds. Do different birds feed at different times of day? Do you only see them at a certain time of year? Do some birds chase other birds away and hog all the food? Use your field guide to see how many birds you can identify. Even without a feeder, you can watch and identify the birds in your yard and neighborhood.

- Gather up your hiking gear (backpack, binoculars, field guide to the birds book, field notebook, pencils, water, and snack) and go for a walk anywhere outside. When you spot a bird, see if you can identify it. Make notes about what kind of bird it is, where and when you saw it, and its color and markings. Also note anything else of interest. What is it doing? What is it eating? What kind of habitat is it in? How many other birds of the same kind are with it? What else can you notice? If the bird sits still long enough and you have a camera, you can photograph it. Or you can make a sketch of it.

- Try pishing. Stand still and make a sound like "psshh, psshh, psshh." Curious birds may gather near you to see what's going on. A little patience may be required.

- If you have a smart phone or tablet, you can get a bird identification app. The Audubon Society has a free app that has bird calls as well as visual identification information (http://www.audubon.org/apps). Another app is called *Peterson Birds — A Field Guide to Birds of North America* and is available for a small fee. With a little research you'll be able to find other apps too. Note that some apps may work only where you have cellphone or wifi connection.

Experiment 11

Finding North

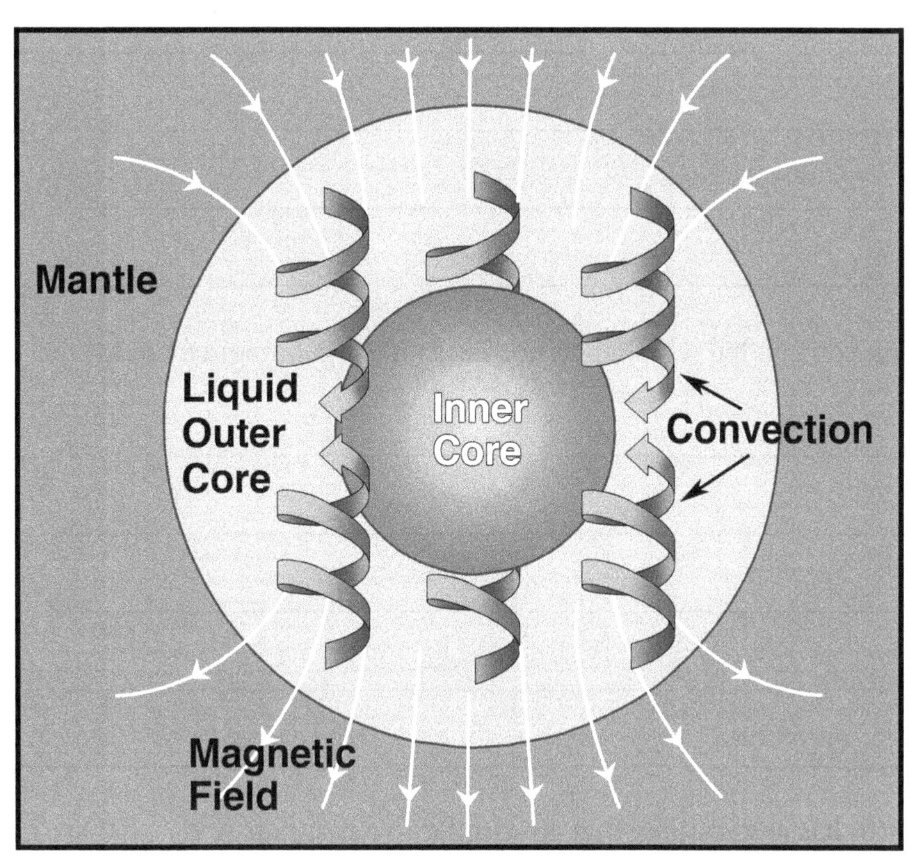

Earth's dynamo—convection in the outer core

Derived from illustration by the US Geological Survey (USGS)

Introduction

Make your own compass and use it to find north!

I. Think About It

❶ How do you think magnets can be useful?

❷ How do you think you can tell whether or not Earth has a geomagnetic field?

❸ Do you think Earth has any features that are similar to a magnet? If so, what are they?

❹ If you were exploring the Arctic, do you think you could find the north end of Earth's axis by using a compass? Why or why not?

❺ When do you think you would want to use a compass and when would you want to use a GPS?

❻ If there were no magnetosphere, do you think life on Earth would be different? Why or why not?

II. Experiment 11: Finding North

Date ___________

Objective

____

____

Hypothesis

____

____

____

Materials

steel needle
bar magnet
piece of cork
tape
medium size bowl
water

EXPERIMENT

❶ Take the bar magnet and slowly stroke the needle against it for about 45 seconds. This will magnetize the needle.

❷ Find an object that a magnet will stick to and test the needle to see if it will stick to that object. If the needle doesn't stick, rub it against the bar magnet for a while more, and then test it again.

❸ Center the magnetized needle on the top surface of the piece of cork and tape it in place.

❹ Pour water into the bowl until it is almost full, and carefully place the cork in the center of the bowl so it is floating, needle-side up. The needle should not be touching the side of the bowl.

❺ Observe what happens. Can you tell which way is north? Why or why not?

❻ In the following box, note your observations.

Observations

❼ In the box in the *Results* section, draw a simple map of the room you are in. Indicate the walls, doorway, and any windows.

❽ Place your map on the table next to your cork and needle compass. Turn the map so that it is in the same orientation as the room. (Match the direction of the walls on your map to the walls in the room.)

In the middle of the map, draw a line that goes in the same direction as the needle of your compass.

❾ Think about the approximate locations of sunrise and sunset in relation to the room you're in. Knowing that the sun rises in the east and sets in the west, mark the approximate locations of east (E) and west (W) on your map.

❿ Can you now tell which end of the needle is pointing north? If so, mark it on your map with the symbol N.

Results

III. Conclusions

Based on your observations, what conclusions can you make about the ease or difficulty of making a compass and finding north? What other conclusions can you draw from your observations?

IV. Why?

In this experiment when you rubbed the steel needle with the bar magnet, you induced a magnetic force that magnetized the needle. A magnet placed close to an object made of a material such as iron can cause that object to become a magnet by a process called induction.

Like a bar magnet, Earth's magnetic field has a north pole and a south pole, and thus Earth is said to have a dipole magnetic field. *Di-* is from Greek and means "two." The needle of a compass points north because the magnetized metallic needle is attracted to Earth's North Magnetic Pole. Earth's magnetic field is referred to as the geomagnetic field, and it surrounds the Earth, going outward from the core at the Magnetic South Pole and in at the Magnetic North Pole.

Scientists think that Earth's magnetic field is electromagnetic. An electromagnet can be built by running an electric current through a wire that is coiled around an iron rod. Although there aren't wires in Earth's core, the geomagnetic field is thought to work in a way that is similar to an electromagnet, with the geomagnetic field being generated deep within Earth in the liquid outer core. Electrical forces are created in the outer core due to convection currents that are caused by variations in the temperature, pressure, and composition of molten iron and nickel. In addition to convection currents, the spin of the Earth causes a spiralling, or coiling, motion in the molten materials. This motion is called the Coriolis force which also aligns the spirals (coils) of molten materials into a north/south orientation. The resulting electrical forces work as an electromagnet, creating the geomagnetic field. Once you induced magnetic force in the needle in this experiment and allowed it to turn freely, the needle was able to "find" the North Magnetic Pole.

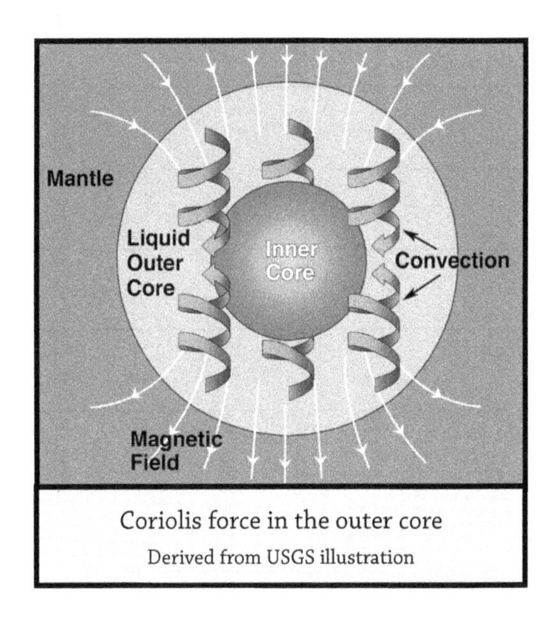

Coriolis force in the outer core
Derived from USGS illustration

V. Just For Fun

Finding treasure!

In Experiment 2 you made a treasure map by using your feet to measure distances to different objects. In this experiment you will make a treasure map by using a compass to show direction along with the number of steps taken.

Materials

small object of your choice for the treasure
compass

Experiment

1. Practice finding North (N), South (S), East (E), and West (W) with your compass. The needle will always be pointing to North. Stand with the compass in front of you, flat in your hand and parallel to the ground. Hold the compass in the same position as you turn in different directions.

   Turn your body until the needle lines up with N. You are now facing North.

   Now turn to the right until your body is lined up with the E on the compass. You are now facing East and the needle will be pointing to your left.

   Turn to the right again until you are lined up with the S on the compass. You are now facing South and the needle will be pointing in the opposite direction, behind you.

   Turn again to your right until you are lined up with the W on the compass. You are now facing West and the needle will be pointing to your right.

   North, South, East, and West are called the four cardinal directions. Each is 90° from the previous one, or 1/4 of a 360° circle. If your compass has degrees shown on its face, the directions will be: N 0° , E 90°, S 180°, and W 270°.

2. This treasure hunt can be done outdoors or indoors. A box is provided in the next section where you can draw your map. First draw an outline of the area where the treasure hunt will take place. Find North and mark it on your map with an arrow in the proper orientation relative to the outline.

❸ Select a place where you would like to hide the treasure.

❹ Pick a starting location. Place an object here or make a mark at this spot. Put the starting point on your map.

❺ Use the compass and the cardinal directions to chart a path to the treasure. On the map write the direction of travel and how many heel-to-toe steps are to be taken between each change of direction. (You can chart your map first and hide the treasure at the end of your route.)

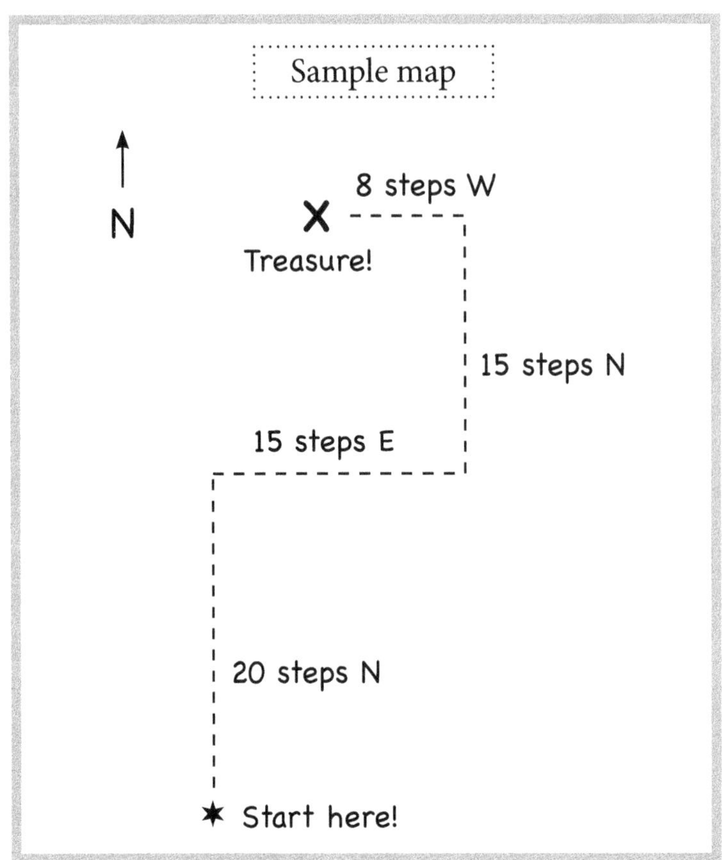

❻ Give the map to a friend and have them use just the map and the compass to find the treasure. You can play this game more than once, refining your map or making the map a bit more complicated each time. Your friend can try making a map for you to follow.

Treasure Map

❼ Record your observations. How well did your map work? Did you have any problems making or using it? If so, how did you solve them? What did you notice about using a compass?

Observations

❽ Look back at Experiment 2, Hidden Treasure. Compare the map you just made with the one you made in Experiment 2. Was one map easier to create than the other? Was one map more accurate than the other? Why?

Experiment 12

Solve One Problem

Cleaning up trash along the Anacostia River, Washington, DC

Courtesy of Gwen Bausmith, US Environmental Protection Agency (US EPA)

Introduction

The solution to a problem starts with a thought!

I. Think About It

❶ Do you think there's a way that Earth's spheres could operate separately from one another? Why or why not?

____

____

____

____

❷ Which of Earth's spheres do you think might be affected by a big blizzard? How and why?

____

____

____

____

❸ If you were an Earth system scientist, how do you think you might look at waste water disposal differently than a chemist? Why?

____

____

____

____

❹ If there is an oil spill in the ocean, which different scientific disciplines do you think might help solve the problem? How?

❺ If you needed to solve an environmental problem in your community, how do you think you might go about looking for new ideas?

❻ What do you think is being done in your community now to help keep Earth's spheres healthy and in balance?

II. Experiment 12: Solve One Problem

Date ____________

Objective ____

____

Hypothesis ____

____

____

Materials

pencil, pen
imagination

Optional

notebook

EXPERIMENT

❶ Spend a day observing your neighborhood. Go outside and watch how the people in your neighborhood interact with their surroundings. Observe how people's interactions affect other people's lives, plant life, and other animal life.

❷ Make a list of all the problems you discover. Include any details you notice. For example, maybe your neighbor is elderly and can't get her trash to the curb, so it piles up, spills over, and then the dogs eat it and scatter it across the street and get sick. Or maybe your neighbors walk their dog but never pick up their dog's waste and, unhappily, other people step in it, tracking it into the local store where it contaminates a cantaloupe that has fallen on the floor on top of it.

❸ Record how many different spheres these problems influence. For example, trash in the yard interacts with: Earth's crust (trash in the dirt), the biosphere (animals eat the trash), the atmosphere (the trash gives off an odor), and possibly the hydrosphere (if it rains and the water becomes contaminated).

Observations of Problems

Problem	Spheres Influenced

❹ Examine the problems and imagine a solution to each problem that you could actually carry out. For example, if your neighbor can't get the trash to the curb in time for the trash pickup, then maybe you could volunteer to move the trash for her. List your solutions below.

Solutions to Problems

Problem	Solution

❺ Now that you have solved the problems, explore how you might expand your solutions to solve the problems for a larger number of people. For example, perhaps you can recruit 5 friends to volunteer so 6 elderly people will have help moving their trash to the curb every week. List how you could expand your solutions to serve more people.

Expanded Solutions

Problem	Solution

❻ Now that you have found a way to expand your solutions to serve a greater number of people, explore how you could turn one of your ideas into a business that would allow you to generate resources and jobs for others while helping to solve more problems. For example, you could start a nonprofit company and get donations from the city government to help recruit more volunteers to help more elderly with their trash. Describe below how your idea could become a business or nonprofit organization.

Results

Summarize your results below.

Results	
Questions	**Ideas**
What was the problem?	
Which of Earth's spheres did the problem influence?	
What was the solution?	
How could you expand the solution to serve more people?	
How could you turn your solution into a business or nonprofit?	
When will you start implementing your solution in real life?	

III. Conclusions

It is said that knowledge is power. Describe below how understanding the science of the Earth — how it is put together and how it works — can help give you the power to solve real-life problems.

IV. Why?

In this experiment you explored how the principles of Earth system science can be applied to solving a problem that exists in your community. By observing how changes in one of Earth's spheres cause changes in the other spheres, we can better understand how human activities affect the balance of the spheres. With this understanding, we can begin to solve problems such as pollution or trash that needs to be put out to be collected.

Careful observation and the collection of data are necessary. What is the problem that is occurring? What is causing the problem? What are the effects of the problem? Once the problem, its causes, and its effects have been identified and analyzed, we can begin to think of solutions. The solutions, in turn, will be analyzed to see what effect they will most likely have on the other spheres. Since the interactions between the spheres is so complicated, outcomes can be uncertain, but analyzing problems and solutions before acting leads to better results.

You may have found that, in a similar way to Earth's spheres, communities can also be thought of as having spheres. Individuals are part of families, families are part of neighborhoods, neighborhoods are part of a larger community, and so on. There are small organizations such as Girl Scouts and Boy Scouts that are part of the national community of scouts. Local governments are part of state governments which are part of national governments. Each part has its own characteristics and actions, and all the parts, large and small, are interrelated and affect each other. Understanding how different groups interact and affect each other is an important step in problem solving.

Because groups and events are so interrelated, it's quite possible that the positive effects of coming up with one seemingly small solution to a problem will spread and have a very positive effect on the larger community.

V. Just For Fun

Now that you have come up with an action for solving a problem, think about how you would let others know about the problem and your idea for a solution. For example, you might create a flyer about the service you are providing for putting out the trash in your neighborhood and then hand out the flyer to your neighbors. You might write a newspaper article or a letter to the editor of your local newspaper telling people about your project, or you might start a blog. If you have a local radio or TV station, you might contact them about doing a news story. You might also make an informational video describing your project and send it to interested people.

In the space below, write your ideas for one or more actions you could take and how you would go about doing them.

More REAL SCIENCE-4-KIDS Books

by Rebecca W. Keller, PhD

Building Blocks Series yearlong study program — each Student Textbook has accompanying Laboratory Notebook, Teacher's Manual, Lesson Plan, Study Notebook, Quizzes, and Graphics Package

Exploring the Building Blocks of Science Book K (Activity Book)
Exploring the Building Blocks of Science Book 1
Exploring the Building Blocks of Science Book 2
Exploring the Building Blocks of Science Book 3
Exploring the Building Blocks of Science Book 4
Exploring the Building Blocks of Science Book 5
Exploring the Building Blocks of Science Book 6
Exploring the Building Blocks of Science Book 7
Exploring the Building Blocks of Science Book 8

Focus Series unit study program — each title has a Student Textbook with accompanying Laboratory Notebook, Teacher's Manual, Lesson Plan, Study Notebook, Quizzes, and Graphics Package

Focus On Elementary Chemistry
Focus On Elementary Biology
Focus On Elementary Physics
Focus On Elementary Geology
Focus On Elementary Astronomy

Focus On Middle School Chemistry
Focus On Middle School Biology
Focus On Middle School Physics
Focus On Middle School Geology
Focus On Middle School Astronomy

Focus On High School Chemistry

Super Simple Science Experiments

21 Super Simple Chemistry Experiments
21 Super Simple Biology Experiments
21 Super Simple Physics Experiments
21 Super Simple Geology Experiments
21 Super Simple Astronomy Experiments
101 Super Simple Science Experiments

Note: A few titles may still be in production.

Gravitas Publications Inc.
www.gravitaspublications.com
www.realscience4kids.com

Printed in the USA
CPSIA information can be obtained
at www.ICGtesting.com
JSHW061210250424
61848JS00002B/16

9 781941 181553